Chugach National Forest

The richest value of wilderness lies...

...not in the days of Daniel Boone, nor even in the present...

...but rather in the future. — Aldo Leopold

Chugach National Forest
LEGACY OF LAND, SEA AND SKY

By Andromeda Romano-Lax

Alaska Natural History Association
Anchorage, Alaska

The publisher thanks the Chugach National Forest for their assistance in developing and reviewing this publication. The Alaska Natural History Association works in partnership with the National Forest Service to further public education and appreciation for the national forests in Alaska.

Author: Andromeda Romano-Lax

Photography: All photos © Ron Niebrugge/wildnatureimages.com except: pages 10-11, © Cliff Reidinger/AlaskaStock.com; page 13, © Cliff Reidinger/AlaskaStock.com; pages 18-19, inset & background & pages 22-23, all photos, Cordova Historical Society; page 30 & 32, © Mark Emery; pages 42-43, background, Chugach National Forest; page 50 & 50-51, background, © Jim Wark/AlaskaStock.com; page 51, inset, © Patrick Endres/AlaskaStock.com; page 53, © Mark Moffit/Minden Pictures; page 60 & 67, US Fish and Wildlife Service; page 68, © Jeff Schultz/AlaskaStock.com; pages 68-69, background, National Park Service; page 69, inset, © Carrie McLain Museum/AlaskaStock.com; pages 70-71, © Randy Brandon/AlaskaStock.com; page 73, © Michael DeYoung/AlaskaStock.com; pages 76-77, © Fred Hirschmann.

Art Direction and Design: Chris Byrd

Illustrations: Kathy Lepley

Map: Courtesy of Chugach National Forest

Editor: Nora L. Deans

Project Coordinator: Lisa Oakley

Forest Service Coordinator: Susan Rutherford

ISBN-13: 978-0-930931-65-0, softcover
ISBN-13: 978-0-930931-85-8, hardcover

© 2007 Alaska Natural History Association. All rights reserved.

750 West Second Avenue, Suite 100
Anchorage, AK 99501

Alaska Natural History Association is a nonprofit educational partner of Alaska's parks, forests and refuges. In addition to publishing books and other materials about Alaska's public lands, Alaska Natural History offers field-based educational programs, teacher trainings and operates visitor center bookstores. The net proceeds from publication sales support educational programs that connect people to the natural and cultural heritage of Alaska's public lands. For more information or to become a supporting member: www.alaskanha.org

Library of Congress Cataloging-in-Publication Data

Romano-Lax, Andromeda, 1970-
  Chugach National Forest : legacy of land, sea, and sky / by Andromeda Romano-Lax.
    p. cm.
  Includes bibliographical references and index.
  ISBN 978-0-930931-85-8 (hardcover : alk. paper) -- ISBN 978-0-930931-65-0 (softcover : alk. paper)
  1. Chugach National Forest (Alaska) I. Title.

SD428.C49R66 2007
333.7509798'3--dc22
                              2006034014

Printed in China on recycled paper using soy-based inks.

Chugach National Forest

Map 1

Introduction 2

Winging North: Copper River Delta 12

Renewal And Return: The Salmon Life Cycle 30

Surrounded By The Sea: Prince William Sound 38

Violent Change: Natural Adaptations 58

Laced With Trails: Kenai Peninsula 66

Afterword 82

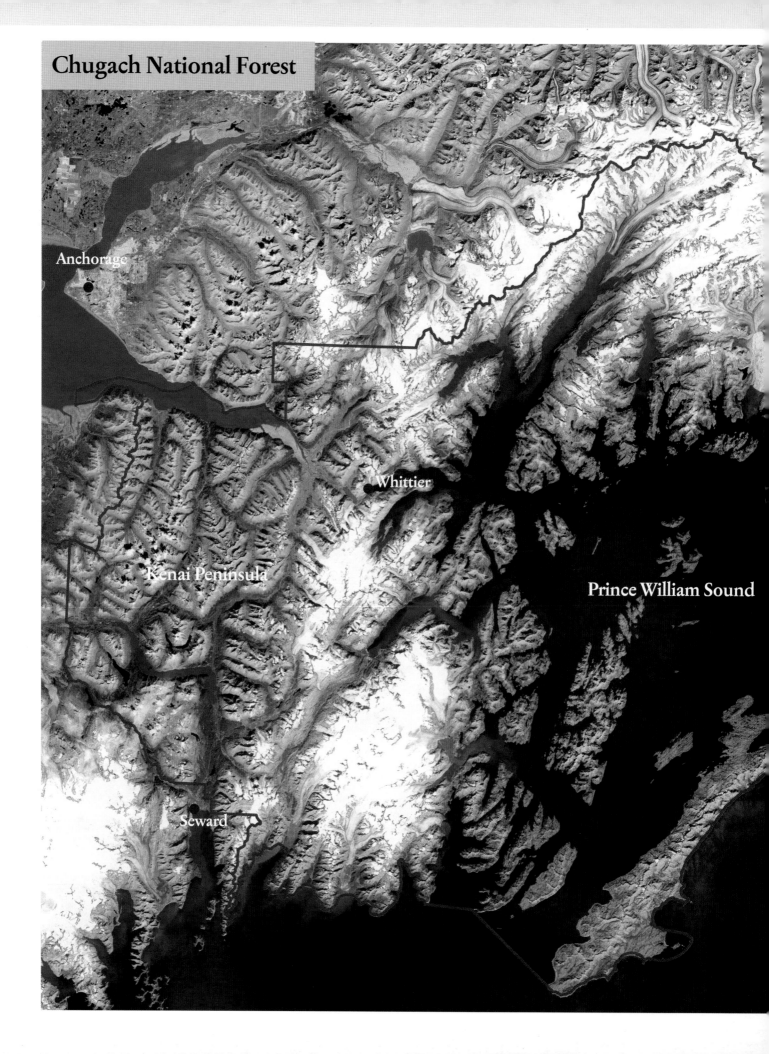

Introduction:
The Great Land and Pinchot's "Greatest Good"

A century ago, parts of Alaska were unrecognizable. Anchorage was not a city, not yet even a collection of tents on muddy Cook Inlet's shore. The population of the state was less than a tenth of today's population. Only forty years had passed since the United States had purchased Alaska from Russia, and statehood (to be gained in 1959) was about as many years away.

Yet in other ways, the Alaska of today had begun to take shape. Resource development had begun in earnest and already, there were concerns about how the land's wild qualities could be conserved. In 1897, big-game hunter Dall DeWeese enjoyed a successful hunt for trophy moose on the Kenai Peninsula. When he returned to hunt again just a year later, he noticed a decrease in game populations and spread word, through the press, that Alaska needed more regulation; other game hunters of the period agreed.

By the following year, 1899, Alaska had thirty-seven salmon canneries. For the first time that year, packers topped the million-case mark. Two years later, drillers at Katalla, east of the Copper River Delta, struck oil. Competing business interests raced to build railroads connecting the coast with the Interior. As the early 1900s progressed, and miners and railroad-builders moved through forested areas, human-caused wildfires changed the landscape.

In addition to all of these changes, the early century saw the birth of a new type of Alaska tourism. Naturalists cruised the Alaska coast, making scientific observations and penning articles and memoirs that would interest many more Americans in visiting this new "district" (as Alaska was called between 1884 and 1912).

Into this time of growth and change strode two great figures: President Theodore Roosevelt and forester Gifford Pinchot, both interested in stemming waste, fighting corporate monopoly, and conserving natural resources for the benefit of all citizens.

Until the late 1800s, federal policy had been to dispose of public lands by giving them away to miners, homesteaders, ranchers and railroad-builders. But as the frontier closed and people realized that America might one day lose its forests and streams to unregulated development, the conservationist ethic was born.

The Great Land and Pinchot's "Greatest Good"

Pinchot, who had worked in the Division of Forestry since 1898, became the newly organized U.S.D.A. Forest Service's first chief in 1905. He is considered not only the father of the Forest Service, but the father of scientific forestry as an American profession. The son and grandson of loggers, Pinchot graduated from Yale University. By the late 1880s, his family—made wealthy by the clearcutting of forested lands to create farmland—became concerned about a future timber famine and supported Gifford's development of new forestry ideas. The search led him to Europe, since the United States had no concept of forest management at the time. The family later endowed a forestry school at Yale.

Both idealistic and pragmatic, Pinchot supported the idea of "the greatest good for the greatest number," a philosophical concept derived from the nineteenth century English writer, Jeremy Bentham. To this concept, Gifford added, "... in the long run."

In contrast with John Muir and other preservationists whose ideas later inspired the national park system, Pinchot believed that public lands were to be used — not withdrawn from use. But he believed they should be managed in a way that would benefit all Americans. For him, that included not only current Americans, but future ones as well. The idea of management was to "insure the permanence of these resources"—a radical concept at a time when, in Pinchot's words, "the common word for our forests was 'inexhaustible.' To waste timber was a virtue and not a crime."

In 1904, Pinchot sent a young forester named William Langille to make a reconnaissance of Southcentral Alaska. Langille considered the forests to be of poor commercial quality. But he also saw that the forest, whatever its long-term value, needed protection. Stands of trees were being cut for use as railroad timbers. Locomotives showered sparks, and work crews left behind slash piles. A large fire on the Kenai Peninsula had been set to get rid of mosquitoes. The trees were destroyed, but the mosquitoes remained.

Langille recommended creation of a reserve, with emphasis on preserving game, including Dall sheep, moose, and the area's last remaining caribou. Alaska's very first forest reserve, on Afognak Island, had been created in 1892 specifically to protect salmon. (The Afognak Fish Culture and Forest Reserve, with its emphasis on fisheries management, would later be incorporated into the Chugach National Forest.) From the start, the national forest system in Alaska would concern itself with much more than trees.

Despite the objections of the General Land Office under commissioner Richard Ballinger, the Chugach National Forest was proclaimed in 1907. Just under five million acres, it extended from the Copper River on the east to the borders of the Kenai Peninsula on the west and inland to the Chugach Mountains.

In the three years since Langille's first visit, "wanton waste" had continued in the region. Along the north shore of Turnagain Arm, the Alaska Central Railroad had set up sawmills and cut three million board feet of timber, which had been left in the woods to decay. Langille advised adding the north shore and other areas to the new forest. To the east, in the Copper River Delta, the Alaska Syndicate, founded by J.P. Morgan and the Guggenheim brothers, was eyeing coal reserves in the public domain to add to its growing monopoly of copper, canneries, and steamship and rail transportation. Allegations of misuse, resource battles, and land-ownership conflicts would be part of the Chugach National Forest's history from its inception.

The legacy of Gifford Pinchot is not an answer to how land should be managed, but a democratic question: "What will the greatest good be?" In the Chugach National Forest, early use emphasized mining and railroad building. In later years, timber waxed and then waned in importance. In recent years, recreation, tourism, and the maintenance of fish and wildlife habitat have become key uses.

Alaskans continue to make a living from the forest, and play in it as well. Visitors come to touch ancient ice, to cast into glacier-fed streams for world-famous salmon, to see majestic scenery. Preserving some areas in their natural, unaltered state has earned recognition as a value of its own. The forest's overall health, clean air and water, and the genetic diversity of resident species have been recognized as deserving the same kind of careful accounting given to colonial Alaska's earliest commercial resources: sea otter pelts, copper, and gold.

At 5.5 million acres and with ninety-six separate watersheds, today's Chugach National Forest is difficult to envision in its vast and diverse entirety. This book adopts the bioregional approach, dividing the forest into three broad geographical regions: Copper River Delta, Prince William Sound, and the Kenai Peninsula. Each of these regions has its own history, its own striking natural features and visitor opportunities. At the same time, the regions share much in common, including abundant salmon populations and dynamic natural processes—two subjects that are covered in separate chapters.

As it celebrates its centennial, the Chugach National Forest has reason to feel proud of its historical Forest Service roots. It has other roots—and other legacies—as well. These legacies are natural and cultural gifts, as well as obligations, passed on to current and future generations. The forest's extensive mantle of glacier ice created a landscape of deep fjords, salmon-rich watersheds, diverse vegetation, and wildlife. Tectonic activity has made—and continues to make—a deep mark on the land, bringing change to an already dynamic ecosystem. The forest's First Nations, including Chugach, Eyak, Ahtna and Dena'ina peoples, created and still maintain a strong sense of stewardship over these lands and waters.

The Great Land and Pinchot's "Greatest Good"

The Great Land and Pinchot's "Greatest Good"

Some legacies have been troubling. Early Russian fur hunters nearly decimated the sea otter population. The 1989 *Exxon Valdez* oil spill severely impacted the coastal forest's marine species. While the Chugach National Forest remains one of the world's most intact temperate rainforests, it faces pressure from other climatic and biological changes, both natural and human-caused.

And yet, the Chugach National Forest has remained remarkably resilient —an ideal place for Pinchot's ideals of democratic participation and "resource permanence" to be tested and, hopefully, achieved. ∎

WINGING NORTH:
Copper River Delta

The largest river that feeds it may bear one mineral's name. But from the air, this vast and famous delta resembles a dozen precious minerals, all polished by the season, the hour, and the light.

Water snakes across the wetlands in ribbons, streaks, and circles that flash silver with the sun, or glow pink at sunset. Dozens of ponds reflect the bright blue sky and the white of advancing clouds, carrying inland the prodigious moisture that marks this coast as temperate rainforest. At higher elevations, the precipitation feeds glaciers—ancient rivers of ice that nearly surround this region.

In May, the delta's low-lying marsh flats turn emerald, glowing with the lush fertility of spring even while nearby mountains remain cloaked in dazzling snow. In drier strips and pockets, shrubs and trees add ribbons of darker green. The colors seem impossibly vibrant, and the air impossibly clear: even from the window of a small airplane, it is possible to spot pairs of white dots in the ponds below. They are the shapes of trumpeter swans, returning to the same nests year after year, for two decades or more.

Even after the sandpipers move on, songbirds like this savannah sparrow fill the Copper River Delta with spring and summer music.

The Copper River Delta stretches across 700,000 acres, draining a watershed of 26,500 square miles, an area about the size of West Virginia. This jewel-toned basin is the largest contiguous wetlands complex on North America's Pacific coast. It is also an ecosystem of astounding biological productivity. Unique within the Chugach National Forest—unique, in fact, among all national forest lands—it is the only one managed by federal law with a priority for fish and wildlife habitat.

The 287-mile long Copper River dominates the watershed. It begins in the Wrangell Mountains, a towering range of sixteen thousand-foot peaks. A tumbling mile-wide river in its upper reaches, it cuts a sharp path through the Chugach Mountains. Falling toward the sea, it is squeezed between monumental walls of ice, including Miles Glacier and Childs Glacier, before widening again. Along its length, the Copper River is fed by over a dozen major tributaries. Its watery lower reaches overlap that of other rivers as well, including the tributaries of the Bering River–Bagley Icefield Complex, an icy realm east of the delta that dwarfs all other glacier complexes outside of Greenland and the polar regions.

As the delta fans out toward the sea, its braided channels split into glittering capillaries and detour around shifting sandbars. Further downstream, wide sloughs swell and contract with the tides. Finally, at the delta's seventy-mile-wide seaward edge, the pulsing water reaches its destination at the Gulf of Alaska. There, it discharges one-third as much sediment as is carried by the Mississippi—a tremendous load, considering the Copper River Delta drains an area only one-fortieth as large. Offshore, large barrier islands form, carved by currents and reshaped by fierce storms. This area of dynamic shallows supports immense populations of marine invertebrates, provides key habitats for salmon at various life stages, and is home to sea mammals, including seals that haul out on sandbars.

COPPER RIVER DELTA

The Copper River Delta Shorebird Festival attracts birders to western sandpiper hotspots like Harney Bay, where up to sixty thousand shorebirds can be seen at one time.

Copper River Delta

Essential Habitats

Picture this maze-like landscape from the air, and you are seeing it as many of its wild visitors do. Beginning each April, the delta becomes a staging ground for sixteen to twenty million migrating birds. Many scientists consider it one of the most essential shorebird habitats in the world. The delta is the largest unit in the Western Hemisphere Shorebird Reserve Network, a system of critical shorebird habitats throughout the Americas. Internationally recognized as a birders' paradise, the delta attracts hundreds of human visitors each May to the Copper River Delta Shorebird Festival.

The birds arrive in waves, even before most of the delta's ponds are ice-free. Some birds stay to nest. Many more stop only to feed and rest before continuing onward to western or Interior Alaska.

First come the swans, ducks and geese, including nearly the world's entire population of dusky Canada Geese, a subspecies that breeds nowhere else. In May, four to six million shorebirds arrive, drawn by food-rich marsh grasses and the extensive mudflats piled up by the sediment-heavy river. The migrants include most of the world's western sandpipers and the Pacific coast's population of dunlins, as well as thirty-four other species of migrating shorebirds. Each sandpiper weighs only an ounce, but the small birds have large appetites. They have flown thousands of miles in just a few days and must be ready to continue their cross-continental trek in a few days more. Using slender beaks, they dip into wet sand with the ease and speed of sewing-machine needles, probing for tiny pink clams, marine worms, and other invertebrates.

Sanderlings, another shorebird in the sandpiper family, race up and down the sandy beach, vigilant for tiny prey uncovered by the waves. Other shorebirds turn over pebbles, pry open molluscs, or submerge their heads in shallow water to retrieve food. While the small shorebirds hunt, they are hunted as well. Merlins, jaegers, peregrine falcons, and eagles all prey on smaller birds. To avoid being

The Chugach National Forest lies at the far northern end of a temperate coastal rainforest extending from northern California to Kodiak Island. It is a landscape of moss-draped trees with high canopies and shady understories. It is also a place of water, falling and flowing in many forms. Raindrops cling to spruce needles. Pendant-shaped waterfalls cascade down rockfaces. Swift-moving streams empty into the cold ocean. Snowflakes compress into glacial ice.

Like their tropical counterparts, temperate rainforests receive at least one hundred inches of precipitation. All rainforests are characterized by thin soils, the presence of large, old trees, and the wildlife that depends on them.

caught, thousands of sandpipers will alight at once, wheeling over the shadows in a flashing cloud of dark wings and white underbellies.

The shorebirds are just one harbinger of warm weather, mirrored beneath the waves by another amazing migrant. All five North American species of Pacific salmon spawn in the Copper River Delta. The most famous is the Copper River red (sockeye), one of the most highly prized fish in the world and, along with the king (chinook) and silver (coho) salmon catch, a vital part of the local economy. The superior taste of Copper River salmon is a direct result of its environment. En route to their spawning grounds, returning salmon do not eat. To survive the trip up the long, cold Copper River, the salmon must have a high fat content. Salmon from smaller watersheds have less fat, and therefore a less rich taste.

Cultural Crossroads

Picture again the aerial view: steep coastal mountains penetrated by a wide, fast-flowing river that widens to a life-filled delta. It's no wonder that this place became home, travel route, and trading corridor for not only one Native people, but many.

As soon as glacial ice began to lose its grip on this rugged landscape about sixteen thousand years ago, people began to move in, following the food sources that were and are plentiful here. Prince William Sound is home to the Chugach, a Pacific Eskimo people. (The Chugach sometimes call themselves the Sugpiaq, and have been referred to by others as Alutiiq). At one time, eight different subgroups of Chugach lived throughout the Sound, with denser populations in the southeast Sound, where food resources were richest.

Upstream, past Childs Glacier, live the Ahtna, an Interior Athabascan tribe. At the Copper River Delta's heart live the Eyak, another Athabascan tribe. Originally from the Interior, the Eyak migrated first to the coast near Yakutat, and later to the Copper River Delta, on the fringe of Chugach territory. Today's national

COPPER RIVER DELTA

A mostly gravel highway leads fifty-two miles beyond Cordova to the spectacular Childs Glacier, which calves directly into the roiling Copper River. The glacier roughly divides an interior climate upstream, and a maritime climate downstream.

Million Dollar Bridge

At the turn of the century, railroad developers all along the coast of southcentral Alaska vied to connect ports like Seward, Valdez, and Cordova with Alaska's mineral-rich Interior. One of the least probable, and ultimately most enduring, projects was the Copper River and Northwestern Railroad, built to transport coal and copper between the Kennecott mines and Cordova.

Railroad entrepreneur Michael Heney purchased the townsite of Eyak, dispossessing the local Eyak Indians. He renamed the location Cordova, created his railroad terminus, and constructed five miles of track before the Alaska Syndicate bought him out, ultimately hiring Heney back to continue construction. Heney died six months before the railroad was finished, in 1911. That April, Cordova celebrated "Copper Day" when the first train of copper ore arrived from the mines and was transferred to a steamship bound for Tacoma, Washington. Over the next twenty-seven years, the Alaska Syndicate became one of the world's top copper exporters, producing a billion tons of copper.

To build a railroad up and across the Copper River required the construction of 129 bridges. One of the most challenging was the Miles Glacier "Million Dollar" bridge, which took two years and considerable money (actually $1.5 million dollars) to build. The cantilevered bridge had to withstand not only floods, but crashing ice from actively calving Miles Glacier, just upstream. The bridge endured many seasons of ice and storm, but was defeated by the 1964 earthquake, which caused one span to drop into the river. A vestige of the empire-building era had finally met its end, eroded not by time, but by the sudden, tectonic forces that continually reshape the Copper River Delta.

Repairs were delayed for years. By the time the bridge was re-opened, the highway meant to cross it and ultimately connect Cordova with the outside world had lost favor. Environmental concerns stopped the highway project. The question of future road development continues to divide the community.

forest includes the homelands and food-gathering places of all of these Native groups, as well as the Kenaitze tribe, a Dena'ina Athabascan people who settled the Kenai Peninsula about a thousand years ago. Southeast of the Copper River Delta, beyond what would become national forest borders, live the Tlingit Indians.

At the center of this cultural crossroads, the Eyak became important middlemen, assisting trade between the coastal Chugach, Interior Ahtna, and coastal Tlingit. The Ahtna are credited with discovering and learning to work the copper found in the watershed's remote headwaters. Through the Eyak, they traded this valuable material, highly prized by Native artisans in Alaska and the Pacific Northwest. Through the Eyak, the Ahtna also traded moose and caribou hides in exchange for sea products.

Introduced diseases, Russian colonization, and early territorial administration all had a devastating effect on Alaska Natives. Despite these tragedies, many resisted displacement from their original territories. Through the Alaska Native Claims Settlement Act in 1971 and Alaska National Interest Lands Conservation Act of 1980, Native groups were granted title to certain public lands within the national forest, which are now held privately by Native corporations. The development choices Native corporations have made in the past and will make in years to come will greatly influence adjacent public lands.

Modern descendants of the region's First Nations continue to live in and near the Chugach National Forest. Their modern stories demonstrate the resiliency of culture and the strong bonds between people and place. One of the communities hardest struck by the 1964 earthquake was Chenega, a Chugach community in southwestern Prince William Sound. Tsunamis unleashed by the earthquake destroyed the old village and claimed the lives of one-third of its residents. Scattered by the natural disaster, determined residents committed to establishing a new Chenega homesite—a

On Kayak Island, naturalist Georg Steller spotted the bird we now call Steller's jay. Recognizing its resemblance to the eastern American blue jay, he wrote, "This bird proved to me that we were really in America."

Copper Kings and the Cunningham Claims

The Alaska Syndicate, founded by J.P. Morgan and the Guggenheim brothers, invested in copper mining, canneries, steamships, goldmining, and other businesses, cornering Alaska resources and transportation to an extent that alarmed Alaska residents—as well as President Roosevelt. In the Copper River Delta, the Alaska Syndicate was particularly interested in developing the Bering River coal fields, a plan foiled when President Roosevelt withdrew all coal and most oil lands in the country from development until Congress could come up with a way to control coal and oil claims.

At the time of this withdrawal, thousands of claims—many of questionable legality—had already been made in the Controller Bay area, in the eastern delta. Rumors circulated that a man named Cunningham, who had filed the claims on behalf of various Washington businessmen, was really a dummy claimant for a large company.

Investigation into the controversy pitted Gifford Pinchot, the Forest Service chief, against Richard Ballinger, who had recently left his post as General Land Office director to practice law. The Cunningham mining claimants had hired Ballinger to defend their case. The next year, Ballinger was named Secretary of the Interior by President Taft. Pinchot continued to research the coal claims and resist the sale of coalfields, publicizing the issue to a sympathetic press. His public discussion of Ballinger's involvement in the scandal brought embarrassment to Taft and resulted in the firing of Pinchot for insubordination, a move that did not surprise the forester.

The scandal led to Ballinger's resignation and the unexpected entry of Theodore Roosevelt, who had retired from politics, as a third-party candidate in the 1912 presidential race. The fracturing of the Republican Party contributed to Democrat Woodrow Wilson's victory. Having sent ripples of political change from Alaska to Washington, D.C., the Cunningham Claims were finally cancelled by the Department of the Interior in 1913.

The Copper River may look like a broad, inviting freshwater route on maps. Actually flying above it, notes historian William Hanable, "one sees not the tempting corridor of the relief map but an icy stream which soon disappears into rank upon rank of snow-capped and jagged mountains." The third view, from foot or boat level, is even more intimidating. "Turbulent silt-laden waters, carrying so much material that all but the strongest man, once submerged, will be carried to the bottom, pass through rock-walled canyons and past constantly calving glaciers to debouch into a mystifying delta which has no certain path to the sea."

Hanable describes the dangers faced by Russian explorers who attempted to penetrate the region—as well as the dangers the Russians themselves imposed upon the Native people of the area. In 1796, a Russian trader named Konstantin Samoilov traveled the river accompanied by Ahtna Indians. One of the Natives dropped Samoilov's tobacco case, made of yellow copper, into the river. The angry Russian told his companions, "Throw him in the rapids. Let him look for my tobacco case." When the Ahtna perished, his companions turned on the Russians, murdering them.

process that took twenty years. The new Chenega is located fourteen miles away from the old village, on a more protected island.

Residents of this new village—like Native residents of many Copper River Delta, Prince William Sound, and Kenai Peninsula communities—maintain many of their ancestor's traditional ways, particularly the gathering of subsistence foods. Though earthquake-caused subsidence of key beaches led to the destruction of rich clam beds, a major subsistence food for the Native people of Tatitlek, residents of that eastern Prince William Sound village continue to harvest many foods from the wild. According to a study in the 1990s, residents of Chenega and Tatitlek averaged 605 and 507 pounds per person of wild foods, more than any other Chugach National Forest community. (Cordova residents harvested an average of 204 pounds per person, about four times as much as residents of Anchorage.) Delicacies that have stood the test of time include salmon, seal, chitons, seaweeds, and more: an edible legacy shared over millennia.

Cordova, Copper Gateway

Cordova deserves a special place in Chugach National Forest history as a town surrounded by national forest, supported by multiple natural resources, and impacted by many of the region's key historical events—including political controversy.

Originally a home to Eyak Indians, the shore of Orca Inlet became home to several salmon canneries beginning in the late 1880s. But another resource, copper, brought the region worldwide fame, as well as a railroad.

In a region where Native Athabascans had long traded copper, rumors of a great copper mountain had intrigued prospectors for years. In 1900, in the Wrangell Mountains near the Chitina River (a tributary of the Copper River), white prospectors stumbled across a great, green cliff of copper ore. Claims were staked and sold, ultimately attracting eastern financiers with enough wealth and clout to

COPPER RIVER DELTA

Jewel-toned salmonberries and blueberries abound on the coast near Cordova.

The vast majority of America's national forests were created well inland, especially in the western United States. The Chugach, like its neighbor the Tongass, is distinctive in that hundreds of its streams and many of its glaciers flow directly into the ocean.

get the riches to market. In 1906, steel magnate J.P. Morgan and the Guggenheim brothers of New York formed the Alaska Syndicate, through which they established the Kennecott Copper Company and financed the building of a railroad from Cordova to the mines.

In 1938, the Kennecott mines closed, leaving behind ghostly ruins in what is now the Wrangell–Saint Elias National Park and Preserve. Even without a railroad, Cordova (population 2,500) has lived on. Today, it remains a town untouched by major highways or by large-scale tourism. Cordova's summer pace is set by the annual return of commercial fishermen and cannery workers, as well as travelers looking for an out-of-the-way destination. ∎

Renewal And Return:
The Salmon Life Cycle

Salmon change their size, shape, and color as they return to spawn in the places where they were born. All five species of salmon develop hooked snouts (called kypes) and jaws, especially visible in males.

Hope is the thing with feathers, poet Emily Dickinson wrote. An even better candidate would be the thing with bright silver scales: a salmon, fresh from years of feeding in the cold ocean, ready to brave obstacles, dodge predators, and struggle for miles upcurrent in order to spawn, all on an empty stomach.

Other animals migrate great distances, but the salmon exerts such energy that it destroys itself in the process, changing color and shape, growing so ragged and battered in the process that some fish look like swimming carcasses. Other animals risk their lives to bear young, but the salmon sacrifices its own life completely. At reproductive journey's end, it dies, achieving not one biological imperative but two. Besides leaving behind the eggs or sperm that will become the next generation, each salmon also dedicates its body to the larger food chain. Nearly everything eats them: bears, wolves, coyotes, and bald eagles. Decomposing carcasses deliver massive quantities of nutrients to the watershed. Salmon rely on the forest, true. But a lesser-known fact is that the forest relies on the nitrogen and other nutrients left behind by dying salmon.

In the 1930s, ethnologists Frederica de Laguna and Kaj Birket-Smith recorded many Chugach legends and stories. Among these was a Chugach story about Raven visiting Sheep Bay, in eastern Prince William Sound. The people there were wealthy in furs and fish, but they didn't yet know how to build a fire. They asked Raven, who had arrived in a canoe, to teach them. He instructed them to give him a silver salmon, which he took the next day and brought to the ocean, shimmering with natural phosphorescence. Raven stirred the salmon in the water, making sparks like fire. Raven came back to land, and the following day he and two men carried the silver salmon tail, and also a small seal stomach full of seal oil. Raven beat the trees with the seal stomach until the oil squirted out. Then he beat the trees with the salmon tail. Raven informed the people that the trees would turn

From about mid-July through September, spawning salmon can be seen in many Chugach National Forest locations. One of the most accessible is the Williwaw Salmon Viewing Platform, about one-and-a-half miles west of the Begich, Boggs Visitor Center, just off the Portage to Whittier road.

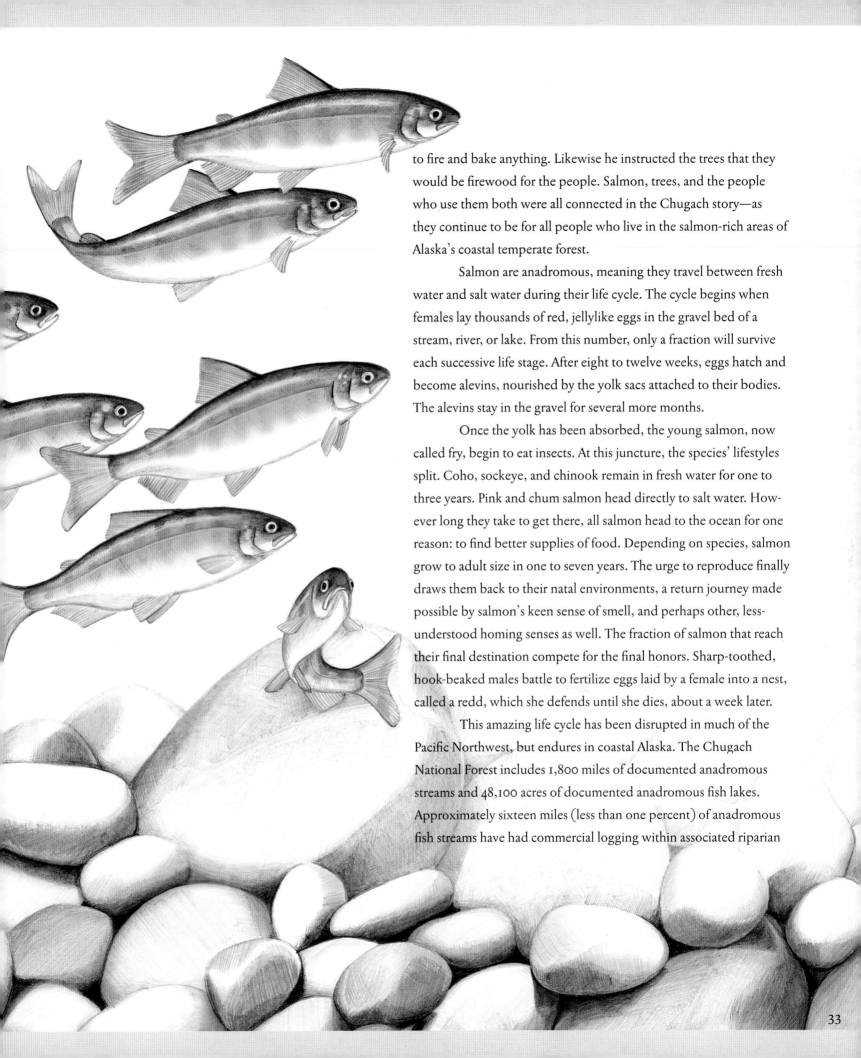

to fire and bake anything. Likewise he instructed the trees that they would be firewood for the people. Salmon, trees, and the people who use them both were all connected in the Chugach story—as they continue to be for all people who live in the salmon-rich areas of Alaska's coastal temperate forest.

Salmon are anadromous, meaning they travel between fresh water and salt water during their life cycle. The cycle begins when females lay thousands of red, jellylike eggs in the gravel bed of a stream, river, or lake. From this number, only a fraction will survive each successive life stage. After eight to twelve weeks, eggs hatch and become alevins, nourished by the yolk sacs attached to their bodies. The alevins stay in the gravel for several more months.

Once the yolk has been absorbed, the young salmon, now called fry, begin to eat insects. At this juncture, the species' lifestyles split. Coho, sockeye, and chinook remain in fresh water for one to three years. Pink and chum salmon head directly to salt water. However long they take to get there, all salmon head to the ocean for one reason: to find better supplies of food. Depending on species, salmon grow to adult size in one to seven years. The urge to reproduce finally draws them back to their natal environments, a return journey made possible by salmon's keen sense of smell, and perhaps other, less-understood homing senses as well. The fraction of salmon that reach their final destination compete for the final honors. Sharp-toothed, hook-beaked males battle to fertilize eggs laid by a female into a nest, called a redd, which she defends until she dies, about a week later.

This amazing life cycle has been disrupted in much of the Pacific Northwest, but endures in coastal Alaska. The Chugach National Forest includes 1,800 miles of documented anadromous streams and 48,100 acres of documented anadromous fish lakes. Approximately sixteen miles (less than one percent) of anadromous fish streams have had commercial logging within associated riparian

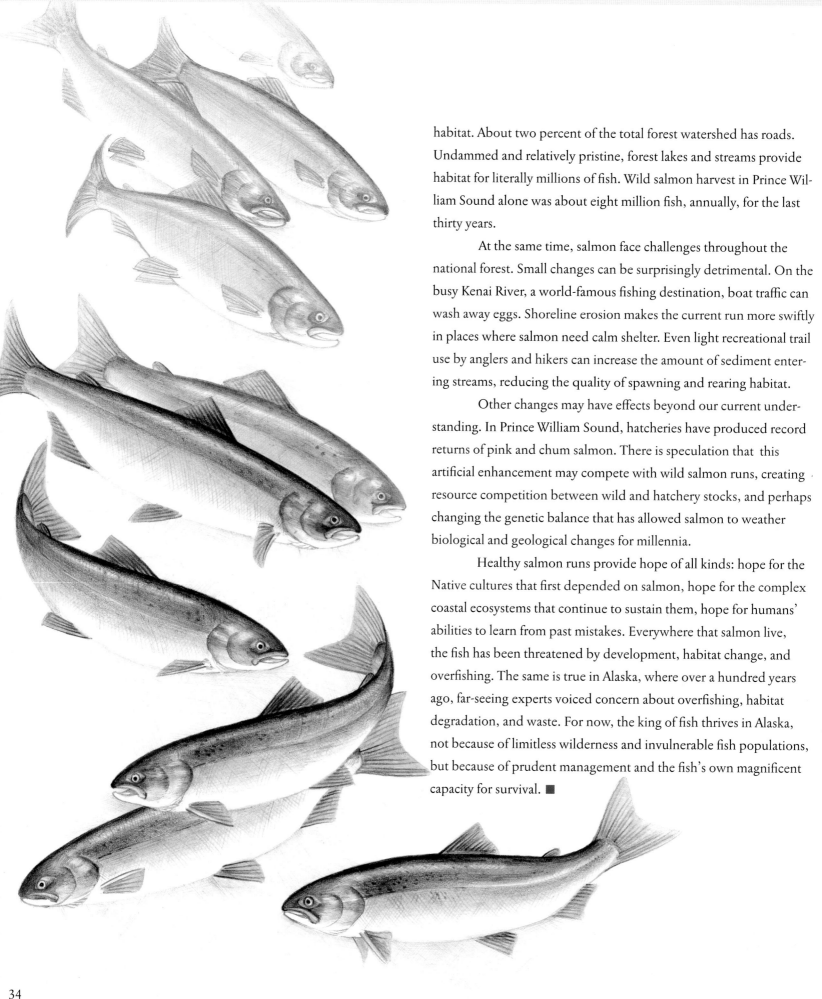

habitat. About two percent of the total forest watershed has roads. Undammed and relatively pristine, forest lakes and streams provide habitat for literally millions of fish. Wild salmon harvest in Prince William Sound alone was about eight million fish, annually, for the last thirty years.

At the same time, salmon face challenges throughout the national forest. Small changes can be surprisingly detrimental. On the busy Kenai River, a world-famous fishing destination, boat traffic can wash away eggs. Shoreline erosion makes the current run more swiftly in places where salmon need calm shelter. Even light recreational trail use by anglers and hikers can increase the amount of sediment entering streams, reducing the quality of spawning and rearing habitat.

Other changes may have effects beyond our current understanding. In Prince William Sound, hatcheries have produced record returns of pink and chum salmon. There is speculation that this artificial enhancement may compete with wild salmon runs, creating resource competition between wild and hatchery stocks, and perhaps changing the genetic balance that has allowed salmon to weather biological and geological changes for millennia.

Healthy salmon runs provide hope of all kinds: hope for the Native cultures that first depended on salmon, hope for the complex coastal ecosystems that continue to sustain them, hope for humans' abilities to learn from past mistakes. Everywhere that salmon live, the fish has been threatened by development, habitat change, and overfishing. The same is true in Alaska, where over a hundred years ago, far-seeing experts voiced concern about overfishing, habitat degradation, and waste. For now, the king of fish thrives in Alaska, not because of limitless wilderness and invulnerable fish populations, but because of prudent management and the fish's own magnificent capacity for survival. ■

35

Surrounded By The Sea:
Prince William Sound

At the turn of the last century, when Alaska's public image was one of gold, copper, and other sources of extractable wealth, one visitor proved prescient. Harry Gannett, Chief Geographer of the U.S. Geological Survey, concluded his report about Alaska this way: "There is one other asset of the Territory not yet enumerated, imponderable, and difficult to appraise, yet one of the chief assets of Alaska, if not the greatest. This is the scenery. Its grandeur is more valuable than the gold or the fish or the timber, for it will never be exhausted. This value, measured by direct returns in money received from tourists, will be enormous; measured by health and pleasure it will be incalculable."

Gannett was one of the many elite guests invited on an 1899 tour of the Alaska coast by east-coast railroad tycoon Edward H. Harriman. The rest of the passenger list of the steamship George W. Elder reads like a "who's who" of the era, including naturalists John Muir, John Burroughs, William Dall, and photographer Edward S. Curtis. Though Harriman was taking the trip for his health and pleasure, he used the opportunity to surround himself with some of the best scientific and artistic minds of the time. They, in turn, were expected to catalogue their observations in the form of scientific reports, artistic works, and popular accounts.

Harriman's "floating university" sailed from Seattle through the Inside Passage, into Prince William Sound, along the Alaska Peninsula and into the Bering Sea. Much of what the expedition members saw as they cruised Prince William Sound, where they spent two weeks of their two-month voyage, can still be seen today.

Imagine, if you will, that trip—one of the earliest ecotourism expeditions—as it crosses from the rough seas of the Gulf of Alaska into the sheltered sound. Rough gray water flattens out into shimmering blue. The dark green ribbon of forested shore reveals itself slowly as the ship passes coves, bays, straits, and steep-walled fjords. Snowy peaks tumble directly into the sea, giving the sound the appearance of a flooded amphitheatre.

At the head of some of the deep fjords stand great walls of glacial ice. Harriman and his fellow travelers approached them incautiously, leaving behind university-inspired names: Columbia Glacier (named for Harriman's alma mater) and College Fiord (with Harvard and Yale glaciers at its forked end); and along the fiord's sides, various male colleges (to the right) and women's colleges (to the left).

Off Barry Arm, the ship dared to enter a narrow passage into yet another ice-choked fjord, to be named later after Harriman himself. John Muir wrote: "The sail up this majestic fjord in the evening

Three to five thousand bald eagles live in Prince William Sound—the equivalent of the entire bald eagle population in the Lower 48. The coast offers them access to salmon and large nesting trees close to the water's edge.

PRINCE WILLIAM SOUND

Mountain goats travel from high alpine, where they evade predators, to the tideline, where they lick the wave-washed rocks for salt. When snows are deep, goats benefit from the presence of densely crowned old-growth coastal trees which intercept snow and make understory forage plants easier to find.

Abundant rain and summer snowmelt feed waterfalls along the forested coastline of Prince William Sound.

sunshine, picturesquely varied glaciers coming successively to view, sweeping from high snowy foundations and discharging their thundering wave-raising icebergs, was, I think, the most exciting experience of the whole trip."

John Burroughs, another famous naturalist who did not always agree with Muir, in this instance echoed the 'other John's' thoughts: "It was one of the most exciting moments of our voyage. ... The scene was wild and rugged in the extreme."

Not only has the shore remained as "wild and rugged" as when Harriman's guests saw it, but most visitors see it today in much the same way, traveling the deeply indented coast in boats of all kinds, from kayaks and state ferries to cruise ships. Every corner of the Chugach National Forest can lay claim to highly scenic areas, but of the forest's three regions, Prince William Sound has the fewest visual signs of human impact. Truly, this is ancient, pristine landscape. Though it's nearly half ice and snow, the remaining acres actually contain a higher percentage of forested land than is found on the Kenai Peninsula or Copper River Delta.

Gannett was a century ahead of his time in predicting people would value something as intangible as scenery. The vast majority of people come to Prince William Sound just to see it—not to do any particular activity, nor to take anything away.

But Gannett and his fellow travelers were very much men of their time, without much concern for effects visitors can have on that magnificent scenery. While Prince William Sound still appears as it did in the Harriman Expedition's day, that scenery, and associated recreational opportunities, require management. The same rugged wildness that makes the Sound a visually spectacular place also concentrates visitors in those areas that are less rugged and easier to reach—along gentler shorelines, in limited campsites and public cabins, and on primitive hiking trails. Recreational use of the forest is expected to increase in years to come. The challenge will be to maintain Prince William Sound's timeless views.

Extreme Weather

Prince William Sound is home not only to extreme weather, but surprising variations in that weather. On average, the coast qualifies as rainforest, with over one hundred inches of annual rainfall. But some communities get more than their share. Valdez receives sixty-five inches of average annual precipitation, Cordova ninety-three inches, and Whittier 197 inches. Less rainy by local standards, Valdez makes up for it in winter, with 303 inches of snow. Cordova receives 118 inches of snow, and Whittier, 250 inches.

The rainiest of all places, not only in the Chugach National Forest, but in the state, is MacLeod Harbor on Montague Island, which in 1976 set state rainfall records with 332.29 inches of precipitation.

Where to look for deep snow? Thompson Pass on the Richardson Highway (outside of national forest boundaries) gets walloped regularly, with 552 inches (forty-six feet) of annual snowfall. In the winter of 1998-1999, staff at the Begich, Boggs Visitor Center recorded 450 inches of snow.

The highest snowpack measured at one time in the state was on Wolverine Glacier, in the Kenai Mountains, southeast of Moose Pass. There, within national forest boundaries, observers measured 356 inches of standing snow accumulated from the 1976-1977 winter.

Portage Valley, just west of Prince William Sound, is notoriously windy. Narrow valley walls channel winds called "williwaws" that have reached 150 miles per hour or more. On Portage Lake, two or three times each summer, boat crewmembers report hurricane-force winds of seventy miles per hour or more.

Unique even by Alaska standards, Whittier was created as a self-contained World War II military community, located on ice-free Passage Canal. To make life easier in this wet, snowy, and windy corner of the sound, the army constructed two of the largest buildings in Alaska at the time: the Buckner and Hodge buildings (later renamed Begich Towers), connected by an underground tunnel. Abandoned by the military in the 1960s, the town's bunker-like lifestyle continues. Most of the town's population continues to live and shop in the fourteen-story Begich Towers.

For years, the only land-based access to the port of Whittier was by train, on tracks that had been blasted through the mountains in record time, during World War II. In 2000, a toll road was completed, using the existing railway tunnel.

Across the sound, Valdez also has a modern, industrial appearance, for different reasons. Founded in 1898 as a road terminus and enriched by gold-rush stampedes, historic Valdez was destroyed by an underwater landslide and tsunamis generated by the 1964 earthquake. The new city was built on safer ground, four miles away. Valdez is best known today as the southern terminus of the trans-Alaska pipeline, bringing oil from the North Slope to tankers in the sound.

Earth in Motion

Every corner of Chugach National Forest can lay claim to an interesting geological past, but here in this land of sharp peaks and retreating glaciers, the Earth's aging process is made especially visible. When the Earth rumbles—as it did catastrophically in 1964, with northeast Prince William Sound at its epicenter—geology takes on a suddenly practical dimension.

The Chugach Mountains that form an arc around Prince William Sound and the Copper River Delta were formed about forty to sixty million years ago when the northward moving Pacific Plate collided with, and subducted beneath, the North American Plate. As the conveyor belt-like Pacific Plate plunged, it dropped off lighter landmasses, called terranes. These terranes included many rock types, including sedimentary rocks that had accumulated at the bottom of a deep seafloor off the California/Oregon coast. The continued movement of the Pacific Plate over time has pushed the Chugach Mountains higher, creating today's range of thirteen thousand-feet peaks. During movement, the rocks were subjected to intense heat and pressure, causing the release of silica-rich fluids that later solidified into veins of quartz. These veins may contain gold and other minerals.

Where tectonic plates meet, as they do across the "Ring of Fire" that stretches from California to Japan, earthquakes are common. The continued northward movement of the Pacific Plate, at about two to three inches a year, causes fifty to one hundred Alaska earthquakes a day. Fortunately, most of these tremors are too small to be felt.

Storm and Ice

The height of the Chugach Mountains is one of the factors, in addition to climate, that makes Prince William Sound so densely glaciated. The mountains catch the constant flow of wet air that blows inland

PRINCE WILLIAM SOUND

Alaska's glaciers have advanced and retreated many times over the past ten thousand years, with most of them retreating since the end of the Little Ice Age. Portage Glacier's retreat began in earnest between 1911 and 1914. As it melted, glacial run-off and rainwater filled up the glacier's basin, creating Portage Lake.

from the Gulf of Alaska, where low-pressure systems generate a steady parade of storms. As the air rises, it cools. Unable to cross the high peaks, the saturated air drops its load of precipitation. Coastal precipitation averages 100 to 160 inches, but in the mountains, this same wet air yields 400 to 800 inches of snow.

Prince William Sound's mantle of ice is astounding, even by Alaska standards. Twenty tidewater glaciers calve into the sound's coastal waters and many more alpine glaciers cling to steep mountainsides.

Glaciers are created where snow never melts completely, allowing old snow to compact over time into multi-year ice. At high altitudes, in the "accumulation zone" made possible by perpetually winter-like conditions, compressed ice changes form, becoming denser, with a different crystalline structure than normal ice. An icefield like the Sargent Icefield in western Prince William Sound spills into numerous valleys, sending rivers of ice that move downhill, succumbing to gravity's pull along the path of least resistance. As these glaciers travel, they erode bedrock and bulldoze debris, shaping the land in their path.

All glaciers are dynamic, simultaneously gaining and losing ice through accumulation and melting. Whether a land-locked glacier grows or shrinks depends on the balance of these two processes. Even a glacier that is shrinking, or losing mass balance, can move forward, appearing to grow simply because gravity is making the living ice flow. Some glaciers reach all the way to sea level. These are the most dynamic of all, with more complicated responses to climate change.

Marine Wildlife

Almost anywhere in Prince William Sound, views of land and sea combine. From a boat, alpine ice and forested uplands are nearly always visible; from nearly any beach or promontory, water sparkles in some—or all—directions. The daily movement of the tides makes the dividing line between these domains even less clear.

Dense gatherings of fish-eating birds are common in Prince William Sound, where over 200 active seabird colonies have been tallied. The cliffs across from Whittier, on the north side of Passage Canal, are home to one of the sound's largest black-legged kittiwake colonies.

Here, in this place of shallow straits, long fjords and protected bays, marine wildlife abounds—and overlaps with the wildlife of tidal zone and forested shore. Bald eagles fly from their spruce-tree nests out and over flooded estuaries, able to spot schooling salmon from a mile away. Orcas, another salmon predator, leave the ocean depths to patrol closer to shore, their tall fins visible above the surface. Sea otters raft in shallows and harbor seals cruise along coves, their round heads emerging just above waterline, with dark eyes fixed on passersby. Even behemoths—migrating gray and humpback whales—find their way between islands and through narrow straits to pay seasonal visits to Prince William Sound waters.

This profusion of marine life has persisted despite natural ecological changes, overhunting, and natural disaster, but it faces challenges still.

At the time of the Harriman Expedition, sea otters were not commonly spotted in Prince William Sound. Turn-of-the-century tourists might have dismissed them as a doomed relic of Alaska's Russian colonial past. From a high of about 150,000 sea otters that ranged from Baja California to Japan, sea otters were nearly wiped out. In 1911, an international hunting ban protected a few sea otter populations of probably no more than one hundred animals each, in Alaska, California, and Russia.

One of these remnant populations lived near Prince William Sound's Montague Island. There, researchers witnessed the return of a species, as 150 sea otters were counted in 1951; 545 in 1959; 2,015 in 1973; and 4,747 in 1985. Though sea otters were hard-hit by the *Exxon Valdez* oil spill (they are currently listed as "recovering"), their current Prince William Sound population of thirteen thousand animals represents a tremendous success relative to their near extinction a hundred years earlier.

Prince William Sound, with its three thousand-plus miles of deeply indented shoreline, makes great sea otter habitat. Though sea

Columbia Glacier, one of Alaska's most-visited tidewater glaciers, has retreated nine miles over the past twenty years. Tidewater glaciers continually flow toward the sea in conveyor-belt fashion, delivering new ice to the glacier face, which melts and calves in contact with seawater. Underwater, the dynamics are even more complex, driven by local variations that may make one tidewater glacier stall in place while another withdraws rapidly towards its alpine origins. Columbia Glacier pulled off its terminal moraine—the underwater sediment pile that anchors a tidewater glacier in place—in the 1980s. Calving accelerated, sending icebergs that threatened oil tankers traveling to and from nearby Valdez Arm. The glacier is expected to retreat another nine or ten miles over the next fifteen to twenty years, creating a deepwater fjord.

Names on the Map

The place-names on a map of the Chugach National Forest attest to the region's international significance. From east to west, the first European name on the map is the oldest. The rocky southern end of Kayak Island was named Cape St. Elias on that Saint's day in 1741 by the first Europeans to set foot in Alaska. Vitus Bering, a Danish navigator sailing under the orders of the Russian Czar, made landfall on the island in order to refill water casks before making the ill-fated return journey to Siberia. Bering reluctantly allowed naturalist Georg Steller to go ashore with a watering party for ten hours—the only time he would spend in this new land. Steller scrambled across the island, gathering natural samples, recording scientific observations, and identifying many plants and animals. Encountering a Native semi-subterranean house that was temporarily unoccupied, Steller took several implements and left behind tobacco, beads, and other gifts. Steller's valiant attempt to make sense of a new land in record time is described in *Where the Sea Breaks its Back* by Corey Ford.

In eastern Prince Willam Sound, the Spanish place names of Valdez, Port Fidalgo, Port Gravina, and Cordova are reminders that the Russians weren't the only eighteenth century European explorers tracing these shores. In 1790, Lt. Salvador Fidalgo named the town of Valdez in honor of a Spanish Naval Officer, Antonio Valdes y Basan, who had planned and financed Fidalgo's voyage. Port Gravina was also named for a Spanish officer. Fidalgo, like Captain James Cook (for whom Cook Inlet is named) and other English explorers of the period, sailed the Alaska coast in pursuit of the fabled Northwest Passage, never found. The Spanish and English both claimed the area but neglected to construct any permanent settlements. Of all the European powers, only the Russians left a cultural legacy in the form of the Russian Orthodox church, to which many Alaska Natives still belong.

Seward, on the national forest's southwest corner, was named for Abraham Lincoln's secretary of state, William Henry Seward, who negotiated the purchase of Alaska from Russia. Critics ridiculed the purchase as "Seward's Folly," and depicted the new territory as a barren icebox—a myth quickly dispelled in decades to come, as prospectors, fishermen, and other entrepreneurs flocked north.

Half of the world's black oystercatcher population lives in Alaska, where the birds are usually spotted singly or in pairs. Kayakers and campers prefer the same low-sloping rocky shorelines as oystercatchers, putting the easily disturbed shorebirds and their coastal nests at risk.

otters feed off the sea floor, where they find mussels, clams, sea urchins, and other invertebrates, they can't dive deeply. That's why sea otters usually are spotted in nearshore areas no deeper than 130 feet. The protected waters of the sound's many bays and fjords provide them with a safe ocean surface on which to groom, rest, and interact with other members of their gregarious species.

Nearshore areas are also the favorite haunts of orcas, especially orcas that hunt harbor seals. Scientists divide orcas into two main groups: transients, and residents. (A third, less-studied group are called "offshores" because they seem to inhabit deeper, offshore waters.) Transient orcas hunt marine mammals. Resident orcas feed on fish and are sighted most frequently near salmon runs—May and early June off the Copper River Delta; and July through September in Prince William Sound.

About three hundred of these top predators live in Prince William Sound and the Kenai Fjords. Highly intelligent and social, the orcas live in matriarch-led pods. They also gather in even larger, multi-pod aggregations of one hundred or more. These "orca conventions," with visiting whales from as far away as southeast Alaska and Kodiak Island, have been observed in lower Knight Island Passage and Montague Strait.

Herring are one of the most important species in the Prince William Sound ecosystem, and yet another example of an animal that depends on inshore areas. Herring spawn in spring, depositing their sticky eggs on intertidal and subtidal vegetation. The 1989 *Exxon Valdez* oil spill occurred at just this sensitive time. Four years later, the herring fishery collapsed. To date, it hasn't recovered. Scientists have studied herring's vulnerability to oil exposure and disease; some speculate that another fish such as pollock may have taken over herring's ecological niche.

Herring is one of Alaska's oldest commercial fisheries and the spring harvest of herring eggs on kelp or hemlock boughs has been

Many seabirds nest in raucous offshore colonies, where they have easy access to fish. But one alcid, the marbled murrelet, prefers to travel far from its food, nesting miles inland, in forest solitude. Specifically, the murrelet requires large, mature trees in structurally complex old-growth coastal forest. In Washington, Oregon, and California, the species is listed as threatened. In Prince William Sound, population declines were noted since the 1970s until about 1990, but the current population of about one hundred thousand birds appears to be stable.

an important subsistence tradition for Natives across Alaska. At every life stage, from egg to adult, the fish becomes food for other local species, including other fish, seabirds, harbor seals, Steller sea lions, and humpback whales. Herring are an emblem of Prince William Sound's interconnectedness, a quality every bit as vital and ultimately valuable as its scenic beauty. ∎

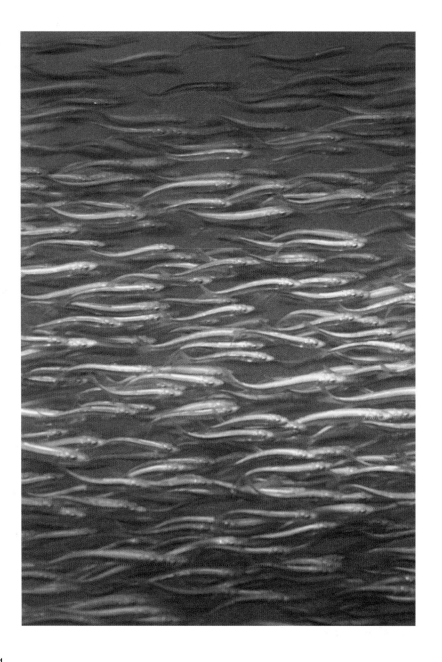

PRINCE WILLIAM SOUND

Though scoured by glaciers and swept by winds, Portage Valley remains a hospitable place for many animals year-round. Bears roam the valley in search of summer berries, while moose shelter from winter storms in the valley's willow and alder thickets.

VIOLENT CHANGE: Natural Adaptations

The dusky Canada goose has a dark chocolate breast and is larger than the lighter breasted common Canada goose. Hunters from Oregon, where the geese overwinter, must pass a bird-identification test before being licensed to shoot common Canada geese.

From the air, it might have looked like a giant, invisible fist had come slamming down on a jigsaw puzzle, forcing the previously aligned pieces to hop and scatter into a jumbled pile. The puzzle pieces were the islands and coastlines of Southcentral Alaska, buckling, lifting, or sinking as the earth shook for about five agonizing minutes on March 27, 1964. Beneath the surface of earth and sea, tectonic forces were unleashed as the North American Plate lurched violently up and over the Pacific Plate.

The 1964 earthquake and even more violent, post-quake tsunamis claimed 131 lives, demolished some communities and scarred many others. Even where towns were not directly threatened, the Earth rearranged itself in unexpected and record-setting ways. With a magnitude of 9.2, the Good Friday earthquake was the strongest ever recorded in North America, and the second most effective at shifting the earth, both horizontally and vertically. Montague Island was uplifted thirty-eight feet in places. Chenega Island was uplifted five feet and shifted fifty-two feet to the south. Underwater, landslides deepened some inlets and bays, while making others more shallow.

The earthquake's legacy has been surprisingly complex, with biological consequences that are still evolving. Before 1964, the Copper River Delta was a vast landscape of shallow waters inundated by high tides. During the earthquake, the spongy delta was picked up and wrung out. Normally, due to coastal plate tectonics, the Copper River Delta subsides—very slowly. The earthquake reversed this process within minutes, leaving the delta six to eleven feet higher than it had been.

Nearly the entire world's population of dusky Canada geese breeds on the Copper River Delta. The 1964 earthquake uplifted their nesting grounds, accelerating the natural succession of the birds' marsh habitat. Nesting sites became higher and drier, putting eggs and birds within the easy reach of more predators, including bears,

In the late 1880s, the Kenai Peninsula was covered mostly with mature spruce and hemlock. Today's mixed deciduous-conifer forests, which contrast with the predominantly conifer coastal forests of Prince William Sound and the Copper River Delta, are partly the result of human-caused fires ignited by railroad builders and early settlers.

wolves, and coyotes. Twenty years after the earthquake, in the mid-1980s to early 1990s, wildlife observers sounded the alarm as the dusky population dropped to about six thousand birds.

The U.S. Forest Service responded by building artificial islands. Still in use, the 370 artificial nest islands function like small boats, anchored to the aquatic vegetation, nearly eliminating mammalian predation. While bald eagle predation on the geese continues to have an effect, the population appears to be stable at about sixteen thousand birds. Stricter hunting regulations in Oregon, where the geese winter, is another important part of protecting the population.

In the short-term, a rapidly depleted dusky Canada geese population represented a disaster for humans and wildlife. On a geological time-scale, the dramatic change was a normal part of the delta ecosystem's life cycle. Every six hundred to nine hundred years, the delta experiences a dramatic uplift, followed by a slow resettling. Dusky Canada geese have evidently survived and persisted despite these recurring events.

Elsewhere in the Chugach National Forest, other long-term cycles attest to the region's dynamic nature. Natural as the cycles are, they can be jarring or threatening to humans; and of course, natural cycles are made more complex with the addition of human-caused variables.

Historically on the Kenai Peninsula, lightning-ignited wildfires haven't been common. From the 1880s to the 1940s, many fires were started by people engaged in mining, railroad construction, and timber harvesting. In modern times, ninety-nine percent of fires are human-caused and small, usually started by campfires. However, prehistorically, widespread fires raged across the Peninsula, about every six hundred years or so, according to charcoal samples from soils.

Fear of future wildfires has increased due to a widespread spruce bark beetle epidemic, which affected over forty percent of the forest since the late 1950s. Land managers must consider human safety and resources in considering how best to deal with large stands of dead trees. But practical issues aside, spruce bark beetle epidemics and resulting wildfires may be just another long-term player in a healthy forest's natural disturbance regime.

At the opposite end of fire, changes in snow and glacial ice prove yet again that the Chugach National Forest is constantly changing. Most glaciers in Southcentral Alaska are retreating or thinning. While human-caused global warming has exacerbated the change, glacial retreat is a natural part of long-term cycles of glacier advance and retreat that have reshaped this land for thousands of years.

Earthquakes have rattled this land; fire has raced across it; rivers of ice have carved deep valleys and retreated, allowing deep fjords to fill with sparkling water. In the wake of all these violent events, natural succession has demonstrated the incredible resilience of both plants and wildlife. Some animals actually thrive in young, early-succession forests. As glaciers retreat, salmon quickly colonize new watersheds. Birds like the dusky Canada geese continue to migrate to the Copper River Delta, despite the region's dramatic changes.

Humankind has both complicated the picture and brought it into sharper focus. Much has been written about the devastation caused by the 1989 *Exxon Valdez* oil spill. As that event recedes in time, one of the surprising aftereffects has been the accumulation of scientific knowledge about fishes, marine invertebrates, sea birds, sea otters, coastal archaeology, and other affected resources.

Ninety years before the *Exxon Valdez* oil spill, the Harriman Expedition produced thirteen volumes of scientific observations that provided a baseline for assessing a century of change along the Alaska coast. That output was unparalleled until today's era of post-spill science, when 1989 became a second baseline against which most of today's fish and wildlife populations are still measured. ■

Laced With Trails:
Kenai Peninsula

In the gold-rush days of the early 1900s, dog sleds and the narrow trails they traversed were a part of daily life, facilitating exploration, hunting, trapping, and the simple but essential delivery of mail. Yet even then, a long trip from the coast into Alaska's frigid heart was considered more than just a grueling commute.

In 1919, politician and musher C.K. Snow wrote about the Iditarod Trail between Seward and Nome: "If you love the grandeur of nature—its canyons, its mountains and its mightiness, and love to feel the thrill of their presence—then take the trip by all means; you will not be disappointed. But if you wish to travel on 'flowery beds of ease' and wish to snooze and dream that you are a special product of higher civilization too finely adjusted for this more strenuous life, then don't. But may God pity you, for you will lose one thing worth living for if you have the opportunity to make this trip and fail to do so."

The Iditarod National Historic Trail, like the many trails that lace the Kenai Peninsula, is the legacy of an era of self-sufficiency and adventure, when the Alaska population swelled with prospectors and other dreamers. Even with gold glittering in their eyes (or after those golden dreams had lost their shine) many early travelers paused to recognize the beauty and majesty of the surrounding landscapes. Today, trails, historic buildings, and rustic cabins connect the Kenai Peninsula to its hard-living, ever-hoping, boom-and-bust past—and to a time before that, when networks of trails were pioneered by Native Alaskans and European explorers, following natural game corridors.

With the scenic Seward Highway and a passenger railway bisecting forest lands, the eastern Kenai Peninsula is highly accessible. Of all the Chugach National Forest lands, these are places where the human footprint—and dogprint—are most noticeable.

Dog-mushing as we recognize it today is a hybrid of Native and European adaptations, reflecting Alaska's diverse heritage. The original Native sled was rigged up in a fan formation, with five to seven dogs individually harnessed. The driver often guided his team along unimproved trail from the front of the sled. In the 1840s, Russians introduced the method of harnessing dogs in single file or in pairs, using a dog leader, and riding or directing the sled from behind. The word "mush" comes from the French-Canadian trapper's

In autumn, paper birch and aspen stands blaze yellow on the Kenai Peninsula, where a mix of deciduous and conifer trees grow.

KENAI PENINSULA

An elusive forest hunter, the lynx preys on snowshoe hares. Both populations peak and crash in paired cycles. Scientists have found that the best lynx-hare habitat occurs where small fires and other natural disturbances create a mixture of vegetation types with an abundance of early successional growth.

The Iditarod Trail

While the route has been traveled since prehistoric times, the Iditarod Trail got its official modern start in 1908, when the U.S. Army's Alaska Road Commission requested a four-man team led by William Goodwin to survey the route from Seward to Nome. Already, gold rushes had transformed areas near the start and end of the trail, Resurrection Creek and Nome, respectively, in the 1890s. But those rushes had waned by the time of the Alaska Road Commission survey.

While a trail to Nome (shorter than the existing route from Valdez, via Fairbanks) was much needed to end that coastal town's winter isolation, the middle parts of any cross-Alaska trail promised to be little traveled unless more gold-rush towns sprung up. As luck would have it, Interior gold was discovered later that same year on a creek near the town of Iditarod, about halfway between Seward and Nome, and just sixty miles southwest of the newly blazed route. The arrival of more than ten thousand prospectors, who settled in Ruby and Iditarod, prompted the final clearing, marking, and improvement of the Seward-to-Nome trail project in the winter of 1910-11.

By the end of the 1920s, air transport challenged dogsled transportation. The Iditarod Trail's final moment of glory came at the dawn of this period, when a relay team of dog drivers transported emergency diphtheria serum to Nome, saving the town. Despite its heroic reputation, the trail languished. Winter users continued to travel parts of its route, and high-profile endurance events like the Iditarod Trail Sled Dog Race called attention to the section of the trail north of Anchorage. In 1978, the trail was declared a National Historic Trail.

Since many early roads and railways were laid over pre-existing trails, parts of the Chugach National Forest portion of the Iditarod National Historic Trail are now unrecognizably modern. But other parts have become wilder, rougher, and less-traveled with time. During the Gold Rush period, the Iditarod Trail was surprisingly well developed, with roadhouses spaced every twenty miles or so, a day's journey apart. Today, much of the trail—originally meant for winter travel—is difficult to find or impassable in summer, due to vegetation and water crossings. Some of the most accessible parts of the trail north of Seward include the twenty-three-mile Johnson Pass Trail, which follows a 1,550-foot pass through lake-dotted alpine backcountry.

The Forest Service plans to construct approximately seventy-six miles of new trail and reconstruct approximately sixty-seven miles of existing trail connecting Seward with Girdwood and Crow Pass.

Tern Lake, at the junction of the Sterling and Seward highways, is an accessible place to see loons and ducks, to listen for songbirds, and to watch the graceful fish-hunting plunges executed by the lake's namesake bird.

word "marché" or "march"—which originally referred to walking or snowshoeing from village to village, with or without dogs.

Backyard Wilderness

The Kenai Peninsula has a reputation as a backyard wilderness within easy driving distance of Anchorage, the state's biggest city. It also has the reputation of being a microcosm of Alaska, with a little of every kind of resource and recreation the rest of the state offers, including world-class fishing, wildlife-viewing, camping, and hunting, not to mention spectacular scenery. The Chugach National Forest is neighbor here to other public lands: Kenai National Wildlife Refuge and Kenai Fjords National Park, as well as the road-accessible communities of Seward, Hope, Cooper Landing, and Moose Pass.

The Kenai Peninsula's most popular multi-day hike is along the Resurrection Pass Trail, a thirty-nine-mile long trail that climbs from 500 to 2,600 feet, passing no roads or towns along the way. The trail fell out of use until 1965, when the Forest Service reopened it to the public. The trail is well known for its no-frills public cabins (reservations required), which were built on the pre-existing sites of "trespass cabins" left by anglers and trappers. Most cabins are spaced from two to seven miles apart.

For an even longer hike, the Resurrection Pass Trail can be linked with the Russian Lakes and Resurrection River trails, for a seven- to ten-day trek between Hope and Seward. However, the area is flood-prone; expect to bushwack and check with rangers for updated information on damaged bridges and trails.

Mountain biking has a surprisingly long history on the Kenai Peninsula. Even during the heyday of dog teams, some budget-minded Iditarod Trail travelers opted to ride (or push) bikes along the frozen winter route. Today's top off-road biking routes include the Resurrection Pass, Crescent Lake, Russian Lakes, and Johnson Pass trails. About four-fifths of the Kenai forest is open for winter-motor-

Forest-covered land makes up only a fifth of Chugach National Forest, but these tree-covered areas are the most species-rich, with about fifty-nine percent of the forest's total species. Freshwater environments come close, with fifty-two percent of the forest's species. But that doesn't make less-speciated habitats any less important overall, especially for some animals that require a variety of places to eat and breed.

In the Chugach National Forest, brown bears require riparian habitats (land and vegetation in and next to water sources, such as streams and lakes) for feeding, refuge, and breeding. But secondarily, they make important use of scrub, tidal estuarine, and alpine habitats, as well as several others, especially as the seasons turn. Wolverines are similarly opportunistic and wide-ranging. The Chugach National Forest provides not only wild land, but wild lands: a diverse patchwork of habitats for animals that need to roam.

ized recreation, such as snowmobiling. In summer, most of the area is closed to off-highway motorized recreation.

Gold Country

Some prospectors used the Resurrection Pass Trail to reach Resurrection Creek, home to one of the Kenai Peninsula's earliest and most significant gold-rushes. Explorers and prospectors had been sniffing around these parts since the mid-1840s, when Peter Doroshin of the Russian-American Company found traces of "color" in the Kenai River and its tributaries.

Legend has it that the first successful find was made by Al King, a loner who discovered gold in 1888—and even managed to keep it secret for a year or two. But as usually happens, word soon got out, and more prospectors came to stake claims on Resurrection and Sixmile Creeks. In 1895, news of one party's find of $40,000 in gold reached Seattle, creating a gold rush to Turnagain Arm. The next year, about three thousand miners arrived. Sluicing of gravels was done by hand during short, cool, mosquito-infested summers—hard work, and not terribly profitable, according to one miner who wrote, "I would not advise anyone to come here with the expectation of finding anything fabulously rich."

A second small gold rush was started in 1898 by miners opting out of the more crowded Klondike Gold Rush, in Canada's Yukon. Hydraulic systems were introduced locally in 1902, and a few miners lucked out, but the region never compared with other Alaska bonanzas. The Turnagain Arm District produced about $1 million in gold between 1895 and 1906 (a fraction of that made in the Klondike Gold Rush, where tens of thousands of miners extracted $300 million in gold). Though some miners hung on, action slowed. Sunrise became a ghost town, and Hope all but emptied. Today, it's home to about two hundred residents.

Small commercial operations still seek gold on the Kenai Peninsula. Some forest streams, including a section of Resurrection-

KENAI PENINSULA

Dall sheep roam high mountainsides, where steep rock provides an escape route from predators, including brown bears. Within forest boundaries, Dall sheep are most often seen as distant, white dots. Outside the forest, excellent viewing is possible on the Seward Highway along Turnagain Arm.

Despite their broad antlers and considerable bulk — 1,400 pounds for males — moose manage to hide among trees. The Kenai Peninsula is home to about 8,000 moose — about 1,000 of them in the Chugach National Forest. The desire to protect moose and moose habitat inspired the creation of federal public lands on the Kenai over a century ago.

Creek that begins at the Resurrection Pass footbridge 4.5 miles from Hope, are open to recreational panning. Gold not only glitters, it sinks—the key to making old-fashioning gold-panning possible. Many Chugach National Forest streams still have gold in them, eroded from mineral-rich mountainsides over thousands of years. The swirling and rocking action of manual gold-panning allows gold to concentrate and settle in the bottom or riffled edge of the pan, while sand and water wash away.

Anglers' Hotspots

While some may look at a map of the Kenai Peninsula and see roads, trails, and gold-rush towns, others see only rivers—including two of the world's most renowned angling streams. The Kenai River is the most heavily fished freshwater river in Alaska. Its popularity is due not only to its famous red, silver, and king salmon runs, but its easy accessibility. Stretches of the milky blue-green river are visible from the Sterling Highway and many services, including Forest Service campgrounds, are located close to prime angling areas.

The Kenai River is a prime example of urban wilderness, where anglers, tourists, and residents live alongside—and frequently cross paths with—moose, bears, eagles, and spawning salmon. A study in 2000 found that eleven to twelve percent of the Kenai River's shoreline and nearshore habitat had been damaged by bank trampling and development. Land and resource agencies, sportfishing groups, and local communities have taken part in numerous bank protection and restoration projects to keep the Kenai healthy and productive.

One of the Kenai's tributaries, the Russian River, is smaller and receives fewer visitors overall, but most of them crowd along the same few miles of lower river, toward the smaller river's confluence with the wider, silty Kenai River. The term "combat fishing" well describes the dense congregation of fishermen that meet at this

In the seventy to one hundred years since placer mining left its mark on Resurrection Creek, nature hadn't managed to erase all the scars. Mine tailings—piles of stream gravel piled to about twenty five feet—were still visible along the creekside and the stream itself was straighter and shorter than it had been. Few trees had managed to spread roots through heaped-up cobble and gravel.

The Forest Service and partners decided to help heal the watershed, which is home to all five Pacific salmon species. Beginning in 2005, crews reconstructed a .8-mile portion of Resurrection Creek, mimicking less-disturbed stretches upstream. Mine tailings were graded to restore the natural floodplain and side channels were reconstructed. Revegetation work on the creek segment, located five miles upstream of Hope, continues with assistance from the Youth Restoration Corps.

The project's most important critics, local fish, have already expressed their initial approval. Migrating king salmon that had previously dashed through the creek were spotted lingering in every one of the restored section's newly created pools after the project's first year, according to a forest hydrologist. Healthier salmon populations will benefit other animals that prey on the fish, including bears, eagles, river otters, marten, and mink.

Alpine tundra vegetation in the Chugach and Kenai mountains is adapted to short, cool growing seasons and thin, rocky soils. During past glaciations, tundra plants survived in isolated ice-free "islands" of terrain within Southcentral Alaska.

confluence—up to one thousand anglers fishing shoulder-to-shoulder during the red salmon run's peak, attempting to cast without crossing a neighbor's lines.

Each river has distinct salmon runs and regulations. The Kenai River is a glacial stream that drains the central Kenai Peninsula. At its headwaters in the Chugach National Forest, past the community of Cooper Landing, it is an icy cold but gently graded, meandering river colored by glacial silt. Beyond forest boundaries it heads west, flowing through Skilak Lake in the Kenai National Wildlife Refuge, past Soldotna, and out into Cook Inlet. The river has two runs of red salmon, with the early run of salmon headed exclusively to the Russian River. Several runs of silver salmon enter the Kenai River in late July through early October. Lake trout are year-round residents of Kenai Lake.

The Russian River, a clear stream that enters the Kenai River near Cooper Landing, between Skilak and Kenai lakes, is open to fly-fishing only for most of the year. Anglers target two different red salmon runs—one beginning about mid-June, and a larger second run beginning in mid-July. Silver salmon also arrive at the Russian-Kenai confluence by late July or early August. A much quieter side of the Russian River is glimpsed by anglers who fish before or after the red salmon season (spring and fall trout fishing are popular) or who hike upstream to fish for freshwater fish species at more remote portions of the Russian River and Russian Lakes. Both the Upper Kenai and Russian rivers are closed to king salmon fishing.

The same places that anglers now frequent were essential fishing areas for the region's Dena'ina Kenaitze people. The banks of the heavily used Russian and Kenai rivers contain fragile but often hidden archaeological remains, including depressions that mark the locations of semi-subterranean houses and food caches. In 1992, the Kenaitze tribe and the Forest Service established the Kenaitze Interpretive Site to preserve, protect, and present important archaeological and natural resources, and to be a gathering place for the sharing of tribal stories and customs.

Humans aren't alone in finding the busy Russian River-Kenai River confluence an exciting place to visit in summer. Brown bears also frequent the area to fish. With bears and people competing for food in close proximity, it's crucial for people to be "bear aware" and to avoid behaviors that put themselves, other people, and bears in harm's way.

Across the Kenai, top threats to bears come from many human sources, including car collisions and kills made in defense of life or property. Other activities that threaten bears include private land subdivision, new road construction, hunting and poaching, and habitat degradation caused by salvage logging of diseased spruce trees. These cumulative impacts have prompted scientists to label Kenai brown bears a "species of concern." Many scientists, agencies, and interested citizens are working together to create strategies for improved co-existence between bears and people. ■

KENAI PENINSULA

The Seward Highway was the first Alaska highway to earn the designation of "All-American Road." South of Girdwood, the 127-mile route bisects the Kenai Peninsula portion of the Chugach National Forest, with easy access to hiking trails, campgrounds, and angling areas.

Chugach National Forest:
Afterword

Aldo Leopold wrote, "The richest value of wilderness lies not in the days of Daniel Boone, nor even in the present but rather in the future."

As the Chugach National Forest celebrates its centennial, forest users look to both past and future to understand what ideas and events shaped the forest, and what direction the forest should take in years ahead.

Even with humanity subtracted from the equation, the Chugach National Forest will continue to evolve; its location in a tectonically and climatically dynamic environment virtually assures that fact. In a changing world, the Chugach's valuable resources—including wilderness recreation opportunities, healthy habitats, and thriving species that are rare elsewhere—will become only more valuable.

New forest management plans are created every ten to fifteen years. Public participation and government accountability have been part of the Forest Service vision since 1905. Since its proclamation, people have felt passionate about the Chugach National Forest. They have expressed strong opinions and endured differences of opinion in order to have a hand in the forest's future—and Gifford Pinchot wouldn't have had it any other way.

The Chugach National Forest belongs to you.

"The vast possibilities of our great future will become realities only if we make ourselves responsible for that future."

GIFFORD PINCHOT, U.S. FOREST SERVICE CHIEF, 1905 TO 1910